YOUR KNOWLEDGE H

Bibliographic information published by the German National Library:

The German National Library lists this publication in the National Bibliography; detailed bibliographic data are available on the Internet at http://dnb.dnb.de .

Imprint:

Copyright © 2010 GRIN Verlag, Open Publishing GmbH
Print and binding: Books on Demand GmbH, Norderstedt Germany
ISBN: 9783668574595

This book at GRIN:

http://www.grin.com/en/e-book/380446/in-vitro-evaluation-of-antagonistic-poten-tials-of-some-yeast-isolates

Prasad Singha, Dharnendra Reang, Mayanglabambam Ranjana Devi

"In-vitro" Evaluation of Antagonistic Potentials of Some Yeast Isolates Against Different Plant Pathogenic Fungi

GRIN Publishing

GRIN - Your knowledge has value

Since its foundation in 1998, GRIN has specialized in publishing academic texts by students, college teachers and other academics as e-book and printed book. The website www.grin.com is an ideal platform for presenting term papers, final papers, scientific essays, dissertations and specialist books.

Visit us on the internet:

http://www.grin.com/

http://www.facebook.com/grincom

http://www.twitter.com/grin_com

In-vitro EVALUATION OF ANTAGONISTIC POTENTIALS OF SOME YEAST ISOLATES AGAINST DIFFERENT PLANT PATHOGENIC FUNGI

Arijit Das Prasad Singha1, Dharnendra Reang, Mayanglambam Ranjana Devi

Abstract:

An experiment was conducted to evaluate the antagonistic potentials of 38 yeast isolates against different post harvest fungal pathogens. Four different plant pathogenic fungi (*Colletotrichum musae, Alternaria solani, Rhizoctonia solani,* and *Fusarium oxysporum f.sp.ciceri*) were tested for their growth and behavior in presence of yeasts in dual culture plates using a modified PDA medium. The potential of antagonism of each yeast isolates were measured by a range of inhibition percentages against the test pathogen in dual plates and according to that measurement, the antagonistic potential was grouped in three categories as- **low (<20% inhibition); medium (21%-50%)**; and **high (>50%)**. According to the measurement high antagonistic potentials were found from Y4 and Y22 against *Colletotrichum musae*; Y7 against *Rhizoctonia solani*; Y49 against *Alternaria solani* but against *Fusarium oxysporum f.sp.ciceri*, no single yeast isolate was found to show more than 50% inhibition in dual culture plates.

Key words: dual culture technique, antagonistic yeasts, bio-control efficiency and plant pathogenic fungi.

Introduction:

Research on the control of plant diseases by phyllosphere applied biological control agents has produced a wealth of informations on a wide range of fungal and bacterial biocontrol agents and their applications for controlling diseases affecting leaves, flowers and fruits. Research into their mode of action and ecological adaptation has provided critical insights that have increased commercial utilization of phyllosphere applied biocontrol agents has proven to be challangeing because the leaf surface presents a relatively hostile environment for introduced microbes with a relative paucity of available nutrients (Beattie and Lindow,1999;Beattie,2002), wide water availability fluxes (Beattie and Lindow,1995), direct exoposure to ultraviolet radiation(Sundin,2002) and infrared radiation and competition with other phyllosphere colonist. Foliar applied biocontrol agents do not have the avoidance strategy of endophytic growth available to pathogens (Beattie and Lindow,1995) and therefore must primarily utilize a tolerance strategy whereby they colonise and survive on the leaf surface or in protected sites that may or may not be the same as the pathogen they are intended to control. Because of the variable effects of the biological and physical environment, the vast majority of phyllosphere applied biocontrol agents research has been focused on glasshouse or storage environment wherethe physical environment is more predictable and stable (Paulitz and Belanger,2000). During the past 20 years several biocontrol agents have been exploited and widely investigated against different postharvest fungal pathogens, like *Aspergillus spp, Botrytis spp, Monilinia spp, Penicilium spp, Rhizopus spp.*(Droby *et al.,* 2002). Many of the early studies aimed at the study of the mode of action and evaluation of the efficacy of some potential biocontrol bacteria, such as, *Brevibaccilus subtilis,* producer of antibiotics (Pusey *et al.,* 1986), however, the application of such bacteria on fruit did not prove to be commercially acceptable. Wilson and Wisniewski(1994) indicated the following characteristics of an ideal antagonist: genetic stability, efficacy

1

at low concentrations and against a wide range of pathogens on various fruit products, simple nutritional requirements, survival in adverse environmental conditions, growth on cheap substrates in fermenters, lack of pathogenicity for the host plant and no production of metabolites potentially toxic to humans, resistance to the most frequently used pesticides and compatibility with other chemical and physical treatments. Yeasts seem to possess a good number of the above mentioned features and during the last few years, research has been made on the selection and study of yeasts (Chalutz and Droby, 1998).

Materials and methods:

Antagonistic potential of various yeast isolates(total 38) were tested through dual culture technique against the pathogens like- *Colletotrichum musae, Rhizoctonia solani, Alternaria solani* and *Fusarium oxysporum f.sp.ciceri.*

For this experiment, 6 mm diameter block of the pathogen from pre grown plates were inoculated on Yeast Potato Dextrose Agar medium supplemented with yeast extract at a distance of 2 cm from the periphery of the sterile plate (9 cm diameter). On the other side of the plate a streak of yeast isolate was made at the same distance as that of the pathogen. It should be noted that both the yeast and the antagonist were of the same age when inoculated. A similar plate but without the yeast isolate was also set up as a control for the experiment. The plates containing the paired cultures were then incubated at 28 ± 1^0C for about 6 days.

Interactions between the pathogen and the yeast isolates were noted when the pathogen had just touched the line where sterile distilled water was inoculated in place of the yeast isolate in the control plate. The pathogen growth was measured and from this observation the percent growth inhibition was calculated as {(Control- Treatment)/Control} ×100. Besides, observation was also made as to whether the yeast isolate was suppressive, competitive or inhibitory towards the test pathogen.

Results and Discussions:

Four different plant pathogenic fungi (*Colletotrichum musae, Alternaria solani, Rhizoctonia solani,* and *Fusarium oxysporum f.sp.ciceri)* were tested for their growth and behavior in presence of yeasts in dual culture plates using a modified PDA medium (with 5.0gm yeast extract/lit.).

1. Against *Colletotrichum musae*

Out of the 38 yeast isolates obtained from different phylloplanes as well as some local yeast formulations, date palm wines and decomposing fruits, five yeast isolates showed better inhibitory effects than rest of 33 isolates on *C musae* in dual culture plates six days after inoculation among which three yeasts were isolated from phylloplanes (Y22, Y26 and Y44) and two (Y2 and Y4) from fermenting solution (wines). Y4 and Y2 though isolated from wines obtained from plants from where the yeasts were grown as contaminant of sugar rich plant extract. Radial growth inhibition was maximum by Y22 to the extent of 71.01% over control followed by Y4 which showed an inhibition upto 53.17%. Y44, Y2 and Y26 showed inhibitory activity to the extent of 42%, 42% and 39.13% respectively after 6 days of inoculation but after that *C musae* grew on the other side of the yeast colonies. The results are presented in the **Table-1.1.**

2

Table 1.1: Growth and behavior of *C musae* against yeast isolates in pure culture:

Serial no	Isolate no.	Nature of radial growth (cm) 6 days after inoculation	Percent inhibition (%)	Nature of radial growth at 10 days after inoculation		
				Dense growth(zone of sporulation) (cm)	Sparse growth (filamentous) (cm)	Total growth (cm)
1	Control	6.9	-	6.0	1.0	7.0
2	Y1	4.4	36.0	5.5	1.5	7.0
3	Y2	4.0	42.0	6.0	1.0	5.0
4	Y3	4.5	34.80	5.2	1.0	6.2
5	Y4	3.3	53.17	5.8	0.6	6.4
6	Y5	5.0	27.50	5.4	1.2	6.6
7	Y6	6.3	8.64	6.0	1.0	7.0
8	Y7	5.9	14.48	5.5	1.2	6.7
9	Y9	6.4	7.24	6.0	1.0	7.0
10	Y10	6.9	0.0	5.7	1.3	7.0
11	Y12	6.3	8.64	5.4	1.5	6.9
12	Y13	4.4	36.0	5.6	1.2	6.8
13	Y14(1)raised	5.7	17.39	6.0	0.9	6.9
14	Y14 dry	4.6	32.75	5.4	1.3	6.7
15	Y15	4.6	32.75	5.4	1.4	6.8
16	Y17	4.4	36.0	5.7	1.3	7.0
17	Y22	2.0	71.01	4.3	1.1	5.4
18	Y26	4.2	39.13	5.7	1.3	7.0
19	Y28	4.4	36.0	5.3	1.3	6.6
20	Y29	4.9	28.98	6.0	0.8	6.8
21	Y33	6.1	11.59	6.0	0.5	6.5
22	Y35	4.6	33.33	5.6	1.0	6.6
23	Y37	4.5	34.78	6.0	0.4	6.4
24	Y40	5.6	18.84	6.0	0.8	6.8
25	Y42	4.4	36.0	5.2	1.2	6.4
26	Y44	4.0	42.0	4.4	1.0	5.4
27	Y47	5.9	14.48	5.6	0.9	6.5
28	Y48	4.4	36.0	5.0	1.4	6.4
29	Y49	6.2	10.14	5.4	0.8	6.2
30	Y51	5.0	27.53	4.8	1.4	6.2
31	Y52	4.3	37.98	5.3	1.4	6.7
32	ML1	4.8	30.43	4.8	1.3	6.1
33	ML2	4.6	33.33	5.0	1.0	6.0
34	ML3	5.2	24.63	5.2	1.4	6.6
35	ML4	5.9	14.48	6.0	1.0	7.0
36	ML5	6.3	8.64	6.0	0.7	6.7
37	ML6	6.1	11.59	6.0	0.9	6.9
38	BY1	4.8	30.43	6.0	0.9	6.9
39	Soil yeast	5.6	18.84	6.0	0.7	6.7

Dual culture plates of yeasts and *C musae* were maintained for few more days to allow the pathogen cover the whole plate and also allowed to sporulate so as to observe the sporulation behavior of the fungus. Full growth on the control plate by *C musae* i.e the pathogen grew to reach the other end of the plate from the point of inoculation, was observed seven days after inoculation giving a total radial growth of 7 cm with 6 cm radial growth having sufficient sporulation indicated by pink colored slimy exudates and confirmed by microscopy and remain same upto 10 days of inoculation. In majority of the cases of interaction between the yeast and C *musae* in dual culture plates; the anthracnose fungus grew slowly and sparsely near to the yeast colony as a streak but in some cases *e.g.* Y22, Y44, Y51, ML1 and a few others, after crossing the yeast colony from underneath, started to grow profusely and reached the margin of the Petri plates as in case of the control plates but with a much reduced zone of sporulation as compared to that of the control plate. The radial growth of *C musae* was totally restricted after 6 days in case of Y2, Y4, Y22, Y26, Y44, Y52 with no sporulation or very few mm of sporulating zone which was obtained in case of Y4 and Y52. Although in case of Y10 there was no radial growth inhibition (0%) of *C musae* upto 6 days but growth suppression was indicated by the ceasation of growth after 6 days. At 10 days after inoculation the zone of sporulation in most of the yeast isolates was almost equal to that of the control plate but in comparison (except Y22, Y44, Y51 and ML1), the sparse growth was much less from 6 days of inoculation due to rapid hyphal dissolution of the test pathogen.Microscopic observation of the inhibited *C musea* hyphae from inhibition zone revealed swelling and development of irregular septation as compared to that of the normal hypahe.

Primary screening for the selection of antagonistic organism against a particular pathogen is normally done by dual culture method using a medium that supports sufficient growth and activity of both the organisms. This practice has been widely adapted in the laboratory. During the present investigation it was found that the potato dextrose medium (PDA) was suitable enough for growth of the test fungus *C musae* but was not suitable enough for sufficient growth of the isolated yeasts. To select a suitable medium for both type of organisms a modified PDA medium was developed using 1% yeast extract into the medium. Yeast extract provided several undefined growth factors to the yeasts for profuse growth and this fact is equally applicable in other filamentous fungi also when tested along with the yeasts in dual culture plates.

The present study on the *in vitro* growth and sporulation inhibition by various yeast isolated from phyllosphere indicate the possibility of using or further developing yeasts as bio-control agent against postharvest disease of banana pathogen *C. musae* and also against other fungal pathogen.

2. Against *Rhizoctonia solani*

The test pathogen *Rhizoctonia solani* is an important sclerotial soil borne fungus causing various diseases of plants like- Sheath blight of rice. After incubation of dual culture plates at 28 ± 1^0C, observations were made at 6 days and 10 days after inoculation respectively. The results were shown in the **Table 1.2**. Maximum inhibition was shown by Y7 isolate (from palm wine) both at 6 days and 10 days after inoculation in dual culture plate and the inhibition percentage was 55.22% (**Plate6**) followed by Y33(46.28%) and Y22(43.28%) at 6 days after inoculation. The isolatesY3, Y9, ML2 and Y44 also showed moderate inhibition at 6 days after inoculation but inhibitory effect did not persist upto 10 days of inoculation.

4

Table 1.2: Growth and behavior of *R solani* against yeast isolates in pure culture:

Serial no	Isolate no.	Nature of radial growth (cm) 6 days after inoculation	Percent inhibition (%)	Formation of sclerotia in the dual culture plate at 6 days after inoculation	Nature of radial growth(cm) at 10 days after inoculation	Formation of sclerotia at 10 days after inoculation
1	Control	6.7	-	Yes	7.0	Yes
2	Y1	6.0	10.44	No	6.3	No
3	Y2	6.4	4.47	No	6.8	No
4	Y3	4.1	38.80	No	5.7	No
5	Y4	5.5	17.91	No	6.4	Yes
6	Y5	6.7	0.0	No	6.7	No
7	Y6	6.7	0.0	No	7.0	No
8	Y7	3.0	55.22	No	3.0	No
9	Y9	4.6	36.26	No	6.7	No
10	Y10	6.7	0.0	No	6.8	Yes
11	Y12	5.0	25.37	No	6.7	Yes
12	Y13	6.3	5.97	No	7.0	Yes
13	Y14(1)ra	6.7	0.0	Yes	6.7	Yes
14	Y14 dry	6.7	0.0	No	6.8	Yes
15	Y15	6.7	0.0	No	6.7	No
16	Y17	6.7	0.0	No	6.8	Yes
17	Y22	3.8	43.28	No	5.3	Yes
18	Y26	6.7	0.0	No	6.7	No
19	Y28	6.2	7.46	No	6.5	No
20	Y29	6.4	4.47	No	6.5	No
21	Y33	3.6	46.28	No	4.7	No
22	Y35	6.6	1.49	No	6.9	Yes
23	Y37	6.5	2.98	Yes	6.6	Yes
24	Y40	4.2	37.31	No	6.0	No
25	Y42	6.7	0.0	Yes	6.8	Yes
26	Y44	4.3	35.82	No	6.0	No
27	Y47	6.7	0.0	No	7.0	No
28	Y48	5.8	13.43	Yes	6.2	Yes
29	Y49	4.5	32.83	No	5.5	No
30	Y51	5.9	11.94	Yes	6.8	Yes
31	Y52	5.4	19.40	No	6.5	Yes
32	ML1	4.5	32.83	No	6.3	No
33	ML2	4.0	40.29	No	6.2	No
34	ML3	4.3	35.82	No	6.3	No
35	ML4	6.7	0.0	No	7.0	No
36	ML5	5.6	16.41	No	6.1	Yes
37	ML6	6.6	1.49	No	6.9	Yes
38	BY1	6.0	10.44	No	6.8	Yes
39	Soil	5.9	11.94	No	6.7	Yes

Out of total 38 yeast isolates, 11 such isolates were found from dual culture plate that completely failed to show any kind of inhibition against *R solani* at 6 days after inoculation. They were Y5, Y6, Y10, Y14 dry, Y14(1) raised, Y15, Y17, Y26, Y42, Y47 and ML4. Except these 11 isolates, rest of the 27 isolates showed more or less inhibitory effect towards growth of the pathogen in dual culture plates at 6 days after inoculation. Out of the 27 yeast isolates (that gave positive result during the test), 11 such isolates were found which inhibited the pathogen growth by a margin of more than 30%. They were Y3, Y7, Y9, Y22, Y33, Y40, Y44, Y49, ML1, ML2 and ML3. But out these 11 yeast isolates only Y7 showed a higher degree of inhibition (55.22%) towards *R solani* growth (well restricted) both at 6 and 10 days after inoculation. Y7 culture was isolated from date palm wine solution and morphologically it was creamish, smooth and fast growing in nature and the direct inhibition might be due to production of some antifungal diffusiable metabolites (Walker *et al.*,1985), competition for space and nutrients (Filonow,1998), volatile compounds (Payne *et al.*, 2000) and due to production of some cell wall degrading enzymes such as beta-1,3-glucanase and mycoparasitism (Wisniewski *et al.*, 1991)or by disturbing the hyphal morphology of the test fungus in the dual culture plate (Suzzi *et al.*, 1994).

The isolates Y5, Y14(1) raised, Y15 and Y26 showed 0% inhibition against the test pathogen in dual culture plates at 6 days after inoculation but interestingly after 10 days of inoculation the radial growth of the pathogen remain unchanged as in earlier observation (6.7cm). It means that however the pathogen extended their growth over/underneath the yeast colony through hyphal strands, it would take some time by this yeast isolates to show the inhibitory effect against the fungal growth and did not allow the pathogen to fully cover the plate. However in some isolates like- Y6, Y13, Y47 and ML4, the fungal growth fully covered the plates after 10 days of inoculation. As *R solani* is a sclerotial fungus, the observation were also made with respect to formation of sclerotia (an overwintering fungal propagule) both at 6 days and 10 days after inoculation and found that in the isolates Y14 dry, Y37, Y42, Y48, Y51 and along with the control plate sclerotial formation was there after 6 days of inoculation in very small amount. But the isolates that showed sclerotia formation at 10 days after inoculation, the amount was very high. However in majority of the yeast isolates there was no sclerotia formation in the dual culture plate.

3. Against *Alternaria solani*

Alternaria solani with wide host range is an important plant pathogen. The dual culture system was followed as described in earlier section 4.5.1. One yeast isolate (Y49) amongst the total 38 which showed maximum inhibition of *A solani* (52.27%) after 6 days of inoculation. The other isolates Y14 (1) raised, Y14 dry, Y26, Y28 were observed to show more than 30% inhibition in dual culture plates. The results are shown in the **Table 1.3.**

Table 1.3: Growth and behavior of *Alternaria solani* against yeast isolates in pure culture:

Serial no	Isolate no.	Nature of radial growth (cm) 6 days after inoculation	Percent inhibition (%)	Nature of radial growth(cm) at 10 days after inoculation
1	Control	4.4	-	6.8
2	Y1	4.1	6.81	5.2
3	Y2	4.1	6.81	5.5
4	Y3	3.9	11.36	4.8
5	Y4	3.7	15.90	4.9
6	Y5	3.6	18.18	6.0
7	Y6	3.6	18.18	6.2
8	Y7	4.0	9.09	5.6

9	Y9	3.5	20.45	5.8
10	Y10	3.5	20.45	5.7
11	Y12	4.3	2.27	4.8
12	Y13	3.3	25.0	4.4
13	Y14(1)raised	2.8	36.36	4.8
14	Y14 dry	2.7	38.63	4.5
15	Y15	3.4	22.72	5.3
16	Y17	4.4	0.0	5.8
17	Y22	3.6	18.18	4.5
18	Y26	2.8	36.36	3.6
19	Y28	3.0	31.81	4.6
20	Y29	4.3	2.27	5.9
21	Y33	3.7	15.90	4.5
22	Y35	4.1	6.81	6.0
23	Y37	4.2	4.54	5.7
24	Y40	4.1	6.81	5.8
25	Y42	3.2	27.27	3.9
26	Y44	3.3	25.0	3.8
27	Y47	3.4	22.72	4.3
28	Y48	3.7	15.90	4.8
29	Y49	2.1	52.27	3.1
30	Y51	3.9	11.36	5.8
31	Y52	3.2	27.27	5.0
32	ML1	4.0	9.09	6.0
33	ML2	3.8	13.63	6.2
34	ML3	4.1	6.81	5.6
35	ML4	3.9	11.36	6.3
36	ML5	4.4	0.0	6.8
37	ML6	4.3	2.27	6.0
38	BY1	4.4	0.0	6.8
39	Soil yeast	3.7	15.90	5.4

Dual culture plate results showed that majority of the yeast isolates presented more or less inhibitory action towards the radial growth of the pathogen except Y17, ML5 and BY1 which were observed with 0% inhibitory effect against *Alternaria solani* in the dual culture plate 6 days after inoculation and only in ML5 and BY1 at 10 days after inoculation the pathogen growth was similar to that of the control plate radially. but rather interestingly in case of Y17 the pathogen could not able to cover the full plate 10 days after inoculation as compared to the control plate where the radial growth of the pathogen was 6.8cm. In that case hyphal growth of pathogen only crosses the yeast colony but behind the yeast growth. The isolate Y49 which was creamish, glistening, fast growing and isolated from banana leaf during summer have shown a higher percentage of inhibition 52.27% 6 days after inoculation and well restricted the pathogen (radial growth only 3.1cm) behind the yeast colony even 10 days after inoculation as compared to the control plate where the test pathogen grew at a radial length of 6.8cm i.e. almost covered the full plate. During the observation some more yeast isolates were found which also have shown a good level of inhibition. They were as Y14(1) raised , Y14 dry, Y26 and Y28 with 36.36, 38.63, 36.36 and 31.81 inhibition percentages respectively against *Alternaria solani*. In this 4 isolates *Alternaria* growth just touched the yeast colony at 10 days after inoculation. In most of the cases the pathogen was able to pass across the yeast colony through hyphal strands at 10 days after inoculation but the radial growth was observed to across the yeast from behind its growth.

4. Against *Fusarium oxysporum f.sp.ciceri*

Fusarium oxysporum f.sp.ciceri is a very important soil borne wilt pathogen that causes wilt diseases in various crops of economic importance. It mainly survives in soil and in the diseased crop debris of previous season left over in the field. Results of *in -vitro* studies are presented in **Table 1.4**. Results revealed that the yeast isolate Y14(1) raised and ML3 inhibited the radial growth of *F oxysporum f.sp.ciceri* to the extent of about 44%. The isolates Y49 and Y28 also showed 41.66% inhibitory effect towards the test fungus. The isolate ML5 showed no inhibition against the test pathogen. Rest of the isolates showed low to moderate (4.16% to 39.58%) inhibition of *Fusarium oxysporum f.sp.ciceri* in dual culture plates.

Table 1.4: Growth and behavior of *Fusarium oxysporum f.sp.ciceri* against yeast isolates in pure culture:

Serial no	Isolate no.	Nature of radial growth (cm) 6 days after inoculation	Percent inhibition (%)	Nature of radial growth(cm) at 10 days after inoculation
1	Control	4.8	-	6.8
2	Y1	2.9	39.58	4.2
3	Y2	3.5	27.08	4.8
4	Y3	3.8	20.83	4.4
5	Y4	3.1	35.41	3.9
6	Y5	3.6	25.0	5.7
7	Y6	3.6	25.0	6.0
8	Y7	3.4	29.16	4.2
9	Y9	3.5	27.08	4.8
10	Y10	3.2	33.33	4.5
11	Y12	3.0	37.50	5.0
12	Y13	3.0	37.50	4.5
13	Y14(1)raised	2.7	43.75	4.7
14	Y14 dry	3.1	35.41	3.9
15	Y15	3.7	22.91	5.5
16	Y17	3.0	37.50	4.8
17	Y22	3.7	22.91	5.8
18	Y26	3.5	27.08	4.7
19	Y28	2.8	41.66	3.6
20	Y29	3.0	37.50	4.5
21	Y33	3.0	37.50	4.0
22	Y35	3.7	22.91	4.8
23	Y37	3.5	27.08	5.2

24	Y40	3.1	35.41	5.1
25	Y42	3.9	18.75	5.0
26	Y44	3.3	31.25	4.7
27	Y47	3.0	27.50	5.0
28	Y48	3.4	29.16	5.2
29	Y49	2.8	41.66	4.4
30	Y51	3.9	18.75	5.5
31	Y52	3.2	33.33	4.6
32	ML1	3.4	29.16	5.4
33	ML2	2.9	39.58	4.8
34	ML3	2.7	43.75	4.7
35	ML4	4.0	16.66	6.1
36	ML5	4.8	0.0	6.0
37	ML6	4.6	4.16	6.8
38	BY1	3.6	25.0	5.7
39	Soil yeast	3.3	31.25	4.9

Dual culture plates presented a wide variety of inhibition percentages like the others but majority of the yeast isolates showed more than 20% inhibition against this wilt fungus in the dual plating except Y42(18.75%), Y51(18.75%), ML4(16.66%) and ML6(4.16%) at 6 days after inoculation. ML5 was the only yeast that showed 0% inhibition against the test fungus. But interestingly, the fungal growth could not be able to extend its growth radially as much as compared to the control plate in ML5 yeast isolate at 10 days after inoculation with 0% inhibition. Whereas in case of ML6 which was observed with a very lower percent of inhibition (4.16%) but was better than ML5, the pathogen grew to a similar length to that of the control plate(6.8cm) after 10 days of inoculation in that yeast isolate. The highest degree of inhibition recorded by Y14(1) raised and ML3 yeast isolates (43.75% in both the cases). Y14(1) raised is a fast growing with creamish, dry and filamentous colony in culture tubes and ML3 is also a fast growing, dry but with whitish chalky colony colour isolated from the mango leaf during the winter. Except those Y14 dry, Y1, Y4, Y10, Y12(1), Y13, Y17, Y29, Y33, Y40, Y44, Y52, ML2 and soil yeast were observed with more than 30% inhibition against the pathogen growth at 6 days after inoculation in the dual culture plates. In majority of the cases it was observed that the further growth of the pathogen was remained either at the edge of the yeast colony or just started to grow beyond the yeast colony. But in some cases pathogen growth took place on the yeast colony and started to grow away from the yeast colony and that isolates were ML4, ML5, Y5, Y6, Y22 and BY1 at 10 days after inoculation.

9

Table 5: Summary of results on antagonistic potentials (*in-vitro*) of yeast isolates against the test pathogens on dual culture plates:

Serial no.	Yeast Isolate no.	Potential of antagonism against			
		Colletotrichum musae	*Rhizoctonia solani*	*Alternaria solani*	*Fusarium oxysporum f.sp.ciceri*
1	Y1	medium	low	low	Medium
2	Y2	medium	low	low	Medium
3	Y3	medium	medium	low	Medium
4	Y4	high	low	low	Medium
5	Y5	medium	nil	low	Medium
6	Y6	low	nil	low	Medium
7	Y7	low	high	low	Medium
8	Y9	low	medium	medium	Medium
9	Y10	nil	nil	medium	Medium
10	Y12	low	medium	low	Medium
11	Y13	medium	low	medium	Medium
12	Y14(1)raised	low	nil	medium	Medium
13	Y14 dry	medium	nil	medium	Medium
14	Y15	medium	nil	medium	Medium
15	Y17	medium	nil	nil	Medium
16	Y22	high	medium	low	Medium
17	Y26	medium	nil	medium	Medium
18	Y28	medium	low	medium	Medium
19	Y29	medium	low	low	Medium
20	Y33	low	medium	low	Medium
21	Y35	medium	low	low	Medium
22	Y37	medium	low	low	Medium
23	Y40	low	medium	low	Medium
24	Y42	medium	nil	medium	Low
25	Y44	medium	medium	medium	Medium
26	Y47	low	nil	medium	Medium
27	Y48	medium	low	low	Medium
28	Y49	low	medium	high	Medium
29	Y51	medium	low	low	Low
30	Y52	medium	low	medium	Medium
31	ML1	medium	medium	low	Medium
32	ML2	medium	medium	low	Medium
33	ML3	medium	medium	low	Medium
34	ML4	low	nil	low	Low
35	ML5	low	low	nil	Nil
36	ML6	low	low	low	Low
37	BY1	medium	low	nil	Medium
38	Soil yeast	low	low	low	Medium

10

In the above table the antagonistic potentials of all 38 yeast isolates against four test pathogens in the dual culture plates were presented. The potential of antagonism of each yeast isolates were measured by a range of inhibition percentages against the test pathogen in dual plates and according to that measurement, the antagonistic potential was grouped in three categories as- **low (<20% inhibition)**; **medium (21%-50%)**; and **high (>50%)**. The According to the measurement high antagonistic potentials were found from Y4 and Y22 against *Colletotrichum musae*; Y7 against *Rhizoctonia solani*; Y49 against *Alternaria solani* but against *Fusarium oxysporum f.sp.ciceri*, no single yeast isolate was found to show more than 50% inhibition in dual culture plates.

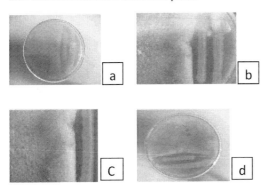

PLATE-6: Interaction between the yeast Y7 and the test pathogen *Rhizoctonia solani* (a-d)

References:

Arras,-G; Fois,-M; D'hallewin,-G. (2007). Enzymatic activity and sequencing of beta-1,3-glucanase gene in biocontrol yeasts. Novel-approaches-for-the-control-of-postharvest-diseases-and-disorders-Proceedings-of the International-Congress,-Bologna,-Italy,-3-5-May,-2007, pp.72-77.

Avis, T. J., Richard, R. and Belanger, R.R. (2002).Mechanism and means of detection of biological activity of Pseudozyma yeasts against plant pathogenic fungi.Federation of European Microbiological Sciences: Yeast Research **2**: 5-8.

Bastiaanse,-H; Bellaire,-L-de-L-de; Lassois,-L; Misson,-C; Jijakli,-M-H. (2010). Integrated control of biological crown rot of banana with Candida oleophila strain O, calcium chloride, sucrose , sorbitols and modified atmosphere packaging. Biological Control. **53**(1): 100-107.

Cao-ShiFeng; Zheng-YongHua; Tang-ShaungShaung; Wang-KaiTuo.(2008). Improved control of anthracnose rot in loquat fruit by a combination treatment of Pichia membranifaciens with $CaCl_2$.International-Journal-of-Food-Microbiology.**126**(1/2): 216-220.

Chalutz, E., Droby, S. (1998). Biological control of postharvest disease. In: Boland, GJ, Kuykendall, LD (eds.), Plant-Microbe Interactions and Biological Control. Marcel Dekker, New York, USA. 157-170.

Chandran-Sandhya; Parameswaran-Binod; Nampoothiri,-K-M; Ashok-Pandey.(2005). Microbial synthesis of chitinase in solid cultures and its potential as a biocontrol agent against phytopathogenic fungus Colletotrichum gloeosporioides.Applied- Biochemistry-and-Biotechnology.**127**(1):1-16.

Dan He, Zheng, X.D., Yin, Y.M., Sun, P. and Zhang, H.Y. (2003). Yeast application for controlling apple post harvest diseases associated with Penicillium expansum. Botanical Bulletin of Academia Sinica**44** :211-216.

Deak,T. (1991). Food borne yeasts.Adv. Appl.Microbiol. **36**.Neidleman, SL, and Laskin, A.I (eds). Academic Press,Inc. New York.

Droby, S. and Chalutz, E. (1994).Mode of action of biocontrol agents of postharvest diseases. In: Wilson, C.L., Wisniewski, M.E. (Eds.), Biological Control of Postharvest Diseases. Theory and Practice.CRC Press, Boca Raton, USA.63-75.

YOUR KNOWLEDGE HAS VALUE